ANNE STEVENSON

The Fiction-Makers

Oxford   New York

OXFORD UNIVERSITY PRESS

Oxford University Press, Walton Street, Oxford OX2 6DP

Oxford New York Toronto
Delhi Bombay Calcutta Madras Karachi
Kuala Lumpur Singapore Hong Kong Tokyo
Nairobi Dar es Salaam Cape Town
Melbourne Auckland

and associated companies in
Beirut Berlin Ibadan Nicosia

Oxford is a trade mark of Oxford University Press

First published 1985 as an Oxford University Press paperback
Reprinted 1985

British Library Cataloguing in Publication Data

Stevenson, Anne
The fiction-makers.—(Contemporary poetry)
I. Title   II. Series
821'.914   PR6069.T45
ISBN 0-19-211972-9

Library of Congress Cataloging in Publication Data

Stevenson, Anne, 1933 Jan. 3
The fiction-makers.
I. Title.
PR6069.T45F5 1985   821'.914   84-29493
ISBN 0-19-211972-9

Printed in Great Britain by
J. W. Arrowsmith Ltd., Bristol

in memoriam
Frances Horovitz 1938–1983

Acknowledgements

Poems from this collection have appeared in the *Times Literary Supplement*, *Stand*, *P N Review*, the *Observer*, *Poetry Durham*, the *Green Book*, *Country Life*, *Outposts*, *Oxford Poetry*, *Other Poetry*, and *Poetry Review*.

'Willow Song', written for Frances Horovitz, appeared in *Tenfold*, Septre Press, 1983; it was also broadcast on the BBC together with 'Where the Animals Go' in 1984. 'Making Poetry' was published as a single poem by Pisces Press in 1983. 'A Dream of Stones', dedicated to Norman Nicholson, is included in *Between Comets*, edited by William Scammell (Taxvs, 1984). Four of these poems are included in *Seagate II, An Anthology of Dundee Writing*, edited by Brenda Shaw (Taxvs, 1984).

A Legacy appeared in a limited edition from Taxvs in 1983, and *Black Grate Poems* were published in a special illustrated edition, with drawings by Annie Newnham, in January 1985 (Inky Parrot Press, Oxford). 'The Road In' is dedicated to Edward Lowbury in whose anthology, *Night Ride and Sunrise*, it appeared in 1978. Special thanks are due to Roger Garfitt whose 'low flying phantoms' I unwittingly borrowed for 'Shale', and to Jon Silkin and Lorna Tracy to whom 'Where the Animals Go' is dedicated.

As always, thanks are due to my husband, Michael Farley.

Contents

Contents

From an Unfinished Poem

. . . The idea of event is horizontal,
the idea of personality, vertical.
Let fiction take root
in the idea of the cross between them.

The mind of the world
is a vast field of crosses.
We pick our way through the cemetery
calling out names and stories.

In the event
the story is foretold,
foremade in the code of its happening.

In the event
the event is sacrificed
to a fiction of its having happened . . .

The Fiction-Makers

We were the wrecked elect,
the ruined few. Youth,
youth, the Café Iruña
and the bullfight set,
looped on Lepanto brandy
but talking 'truth'—
Hem, the 4 a.m. wisecrack,
the hard way in,
that story we were all at the end of
and couldn't begin—
we thought we were living now,
but we were living then.

Sanctified Pound, a knot
of nerves in his fist,
squeezing the Goddamn iamb
out of our verse,
making it new in his
archeological plot—
to maintain 'the sublime'
in the factive? Couldn't be done.
Something went wrong
with 'new' in the Pisan pen.
He thought he was making now,
but he was making then.

Virginia, Vanessa,
a teapot, a Fitzroy fuss,
'Semen?' asks Lytton,
eyeing a smudge on a dress.
How to educate England
and keep a correct address
on the path to the river through
Auschwitz? Belsen?
Auden and Isherwood
stalking glad boys in Berlin—
they thought they were suffering now,
but they were suffering then.

Out of pink-cheeked Cwmdonkin,
Dylan with his Soho grin.
Planted in the fiercest of flames,
gold ash on a stem.
When Henry jumped out of his joke,
Mr. Bones sat in.
Even you, with your breakable heart
in your ruined skin,
those poems all written
that have to be you, dear friend,
you guessed you were dying now,
but you were dying then.

Here is a table with glasses,
ribbed cages tipped back,
or turned on a hinge to each other
to talk, to talk,
mouths that are drinking or smiling
or quoting some book,
or laughing out laughter as candletongues
lick at the dark—
so bright in this fiction
forever becoming its end,
we think we are laughing now,
but we are laughing then.

Waving to Elizabeth

For Elizabeth Bishop (d. 1979)

For mapmakers' reasons, the transcontinental air routes
must have been diverted today, and Sunderland's stratosphere
is being webbed over by shiny almost invisible spider jets
creeping with deliberate intention across the skin-like air,
each suspended from the chalky silk of its passing. Thready at
    first,
as if written by two, or four, fine felt nibs, the lines become
    cloudy
as the planes cease to need them. In freedom they dissolve. Just
as close observation dissipates in the wind of theory.

Eight or nine of them now, and all writing at once,
rising from the south on slow rails, slow arcs, an armillary
prevented by necessity from completing its evidence,
but unravelling instead in soft powdery stripes, which seem
    to be
the only clouds there are between what's simply here as park,
house, roof, road, car, etc. and the wide long view
they must have of us there, if they bother to look.
They have taken so much of us up with them, too—

money and newspapers, meals, toilets, old films, hot coffee—
yet the miles between us, though measurable, seem unreal.
I have to think, 'Here it is, June 19th, 1983.
I'm waving from a waste patch by the Thornhill School.'
As perhaps you think back from your trip through the cosmos,
'Here where I love it is no time at all. The geography
looks wonderful. This high smooth sea's more quiet
    than the map is,
though the map, relieved of mapmakers, looks imprisoned
    and free.'

The Blue Pool

After the painting by Augustus John

It is high summer by the blue pool.
Our heroine has left the safe house of her book
to repose on one arm in the shape of a girl
in this hungry man's painting of a blue pool
with a creamy shelf of dry mountains around it
and a tawny bronze tint to the white reflections.
This could be a hot day near a flooded quarry.
The flowing green dress, the moth-starred jerkin,
the dark bobbed hair are all parts of a story,
but its title and substance have no importance.

What *is* important is the book's colour, which is yellow.
In a child's picture the sun would be this colour,
and the view would fall away from the sun
like a symmetrical tent from a pole.
But in this adult landscape the sun's understood,
it's the undefined source of the light,
so the girl and her book have a moony existence—
as the mind indeed has when it ceases to see *as*
but returns pure reflections from the blue or brown
pools of its seeing. The girl could not possibly be reading.

She herself is the quick of the paint's observation
which allows her so sweetly to float out away from herself,
where for once she is perfectly happy, perfectly whole,
though she still keeps her finger on her place in the book.
The enamelled bright pages must have something in them,
in a minute she's sure to remember . . .
She is young, too, and wishes the painter were with her.
Soon she'll slip round the blank fearful page of his easel
to look at herself. Will the painting be like her?
She will think, she thinks, of something intelligent to say.

The Television and the Nightingale

The Lebanon is on fire in this glassy rectangle.
It shines, and shining, chooses to be true.
A war. A war. And for relief, bad marriage
in which an actress who's a woman acting like
an actress (acting) lends me misery, pure
misery, in a guilt frame like a mirror. Good.

This woman could be I if I could see
into the rectangle (O tangled wreck) of me.

And now some stern ambassador of worldliness
is talking to her through his slim moustache.
But you, also, are speaking through her crash of
fist through door-glass. 'Nightingale',
you say. (Whose rage, whose need, whose fist *is* this?)
'A nightingale is singing in the park outside.'

In the glass that won't be broken, glass is broken.
In the room that won't be glass, a shard of glass
slips in and lodges in my name like ice.

Love, you have brought your nightingale as flower to fish.
Bribes will not better this case, wrong, wrong from the first.
She hates you. I hate you. Why are you spoiling the story?
(When I know it's the story that's spoiling your nightingale.)
I'm hating her and fighting me, fighting a frozen face
bereft of nightingale, lost to its saving voice.

Where the Animals Go

The Beasts in Eden
cradle the returning souls of earth's animals.

The horse, limp cargo, craned down to the terrible quay,
is butchered into the heaven of his own hoofed kind.

The retriever mangled on the motorway, the shot
Alsatian by the sheepfold, the mutilated black-faced sheep—
they rise like steam, like cumulus, crowding in together,
each into the haunches of its archetype.

The drowned vole, the pheasant brought down with his fires,
the kitten in the jacket of its panicking fleas,
flying souls, furred, feathered, scaled, shelled, streaming
upward, upward through the wide thoughtless rose empyrean.

God absorbs them neatly in his green teeming cells.

There, sexed as here, they're without hurt or fear.
Heaven is honeycombed with their arrivals and entries.
Two of each Butterfly. Two of each Beetle.
A great Cowness sways on her full uddered way.
All kinds of Cat watch over the hive like churches.
Their pricked ears, pinnacles. Their gold eyes, windows.

Ailanthus with Ghosts

Their veins were white
but they still hung on with weak hands.
Their wings were dry,
but they still waved back at the wind,
'Just one more day. Just one more day.'

Then, very quietly, Monday night,
a frost-gun shot them away.
Imagine the whole population of Heaven
(Heaven was the name of the tree)
Falling down simultaneously,

dazzling the root in its bed.
'What a beautiful star,' Heaven said,
Or might have if heavens could notice
the difference it makes to an earth
to be thatched with ghosts.

Shale

that comes to pieces in your hand
like stale biscuit; birth book
how many million years
left out in the rain. Break back

the pages, the flaking pages,
to reveal our own hairline habitations,
the airless museum in which we're
still chained into that still ocean,

while all this burly and stirring water—
motion in monotonous repetition—
washes with silt our Jarassic numbness,
the shelves of ourselves to which we will not return.

Bedded in shale, in its negative evidence,
this Venus shell is small as maybe she was.
The fan-shaped tracery of vertical ridges
could be fine-spread, radiant hair,

or proof of what we take to be
her temper—hot sluttishness loosened
by accident into cold mudslide,
preserving a hated symmetry, a hated elegance.

There is so little sheltered, kept, little
and frail, broken in excavation, half
buried, half broken, poor real child in the boulder
that finds the right shape of its mind

only at the moment of disintegration.
And yet—this clear cuneiform in rock;
this sea urchin humping its flower under
'low flying phantoms' . . . this flowing anemone.

Gannets Diving

The sea is dark
by virtue of its white lips;
the gannets, white,
by virtue of their dark wings.

Gannet into sea.

Cross the white bolt
with the dark bride.

Act of your name, Lord,
though it does not appear so
to you in the speared fish.

Taking Down the Christmas Tree

Twelve days, twelve nights, and another Christmas is lost.
The dead spruce is dismantled, its costume of
shining but each year's more shabbily resurrected angels
      is undone at last,
is returned globe by globe, horse by gaudy little
brocade horse, star, and incandescent sequined dove
      to the recurring past.

And now the room is being groomed and put together
by this same brisk hand and snorting machine that helplessly
prepared its dazzle. The children, who came
      home like squally weather,
human in the storm of everything they expected to be
marvellous, would not call this 'Christmas'. But it's the same
      thing as the other,

this sweeping up, implicit in the dangerous promise
of celebration we, almost without thinking, risk
at the winter solstice, bringing the cold forest
      into the warm house.
This harvest of spiny unlooked for needles is almost
seed; and the victim, decked out for worship, is at last
      lost, sacrificed,

so the rite of birth can be death's from the beginning.
What can I say to this sixteen-year-old who packs up
his music patiently and crams for Oxford? Or to Maya at ten
      who minds not winning
at Scrabble but is brave about it? 'The putting up
and taking down of a tree are in time one action,
      as the spring

is one action with autumn, and winter with summer?'
I breathe thanks, instead, to the vacuum cleaner which prevents
too much metaphysics and mother-tears. But the tree
      will not all disappear.
In April I'll trace it again, I know, beating these carpets.
Or, lazing in August and stung by what should be a bee,
      I'll pull spruce from my hair.

An April Epithalamium

For John and Anne Hughes

I meant to write a poem upon your wedding
Full of advice and hidden, deeper meaning.
Alas, my life has locked me out of language.
My sons skulk in their slum of drums and dinner;
Distracting wars break out on distant islands;
Rooms, uncurled in sunlight, cry for cleaning.

I'll hum some thoughts in rhythm while I'm cleaning.
Marriage, you know, is not a life-long wedding,
A launching of moony pairs to pearly islands
Where love, like light, illuminates pure meaning.
For just when truth's in sight, it's time for dinner.
Or lust (thank God) corrupts pure love of language.

Love is, of course, its appetites and language.
Nothing could be more human or more cleaning.
It seems a shame to have to think of dinner
And all the ephemeral trappings of a wedding
When what you pay for seems to cost its meaning.
Are canapes and cake somehow small islands,

Symbols in champagne of all the islands
We try to join together through our language?
John Donne was very sure about his meaning:
No John or Anne's an island. For the cleaning
up and linking up of feelings, a wedding's
A kind of causeway, then—like dinner.

O.K. A man (not John) could wed his dinner.
God help him to imagine lusty islands
Where sun and sea began life with a wedding,
Begetting—not with greedy need of language—
Greenness and creatures (winds to do the cleaning)
That ring-a-rosy in a dance of meaning,

Without which *love* might be the only meaning.
I mean, of course, that *love* and *war* and *dinner*
And *politics* and *literature* and *cleaning*
Are only words, flat atlases of islands.
While, with our mouths, we caterwaul a language,
Our eyes and bodies meet and make their wedding.

But look! I've spoiled your wedding with a meaning,
Tried to spice up with language good plain dinner.
Off to your island now! Leave me my cleaning.

Mother and Son

The woman in romantic confusion
gives birth to the Old Stone Age.
Its coiled spirit calls her
from the peaty dark of the shaft.

From her ten-thousand-year-old window
she would marvel at this slippery wonder,
but she feels too whitely
the pull of his paleolithic gravity.

Down, down he would have her
to drown in his purposeless mouth.
And she, who must haul him up,
emptied at the crumbling surface.

In his canyon he hungers.
From his shell pocked strata he
howls into the tiny blue
aperture of her eye.

A pillow of flesh. A pillar of bone.
A fountain. Up, up,
she draws him, though the circles
of mica and iron.

He invents the hammer, the spear, the arrow.
With a liquid flick of her awl
she secures him in
speech and greediness.

Adam, his evil, Able, his enemy,
he is Thor's boy, Jove's boy
before he is Greek and articulate.

Civilized by the rite of the
codified arena, cruelty and pleasure
war for the power of his story.

But his heart sickens.
He justifies himself in self-blame.
'Mary, mother Mary,
lift me from the gulf of my affliction!'

So, through the dark ages
she is his shawl and beacon.

But now light, or twilight,
or darkness the colour of lightness
shimmers like a visible echo
in his cave of weight—
his mind, *his* music, *his*
electricity of language!

All the world's his drum,
and to use it, use it,
beating at the white stretched tegument—
that is his freedom and torment.

Who keeps him in his glory of
triumphant enlightenment?
With immaculate fingernails
he flicks off the dust of his passage.

The woman in her burrow
looks up to him, boasting and fussing.
But her mind broods lonely
on the sands by a tidal river.

Listen! she calls to him,
to his bright nidicolous cockpit,
Listen! to his feathering will
that ravages tomorrow for the New Stone Age.

nidicolous   as of unfledged birds

Hands

Made up in death as never in life,
mother's face was a mask
set in museum satin.

But her hands. In her hands,
resting not crossing on her paisley dress
(deep combs of her pores,

her windfall palms, familiar routes
on maps not entirely hers
in those stifling flowers), lay

a great many shards of lost hours
with her growing children. As when,
tossing my bike

on the greypainted backyard stairs,
I pitched myself up, through the screen door
arguing with my sister, 'Me? Marry?

Never! Unless I can marry a genius.'
I was in love with Mr. Wullover,
a pianist.

Mother's hands moved *staccato* on a fat ham
she was pricking with cloves.
'You'll be lucky, I'd say, to marry a kind man.'

I was aghast.
If you couldn't *be* a genius, at least
you could marry one. How else would you last?

My sister was conspiring to marry her violin teacher.
Why shouldn't I marry a piano
in Mr. Wullover?

As it turned out, Mr. Wullover died
ten years before my mother.
Suicide on the eve of his wedding,O, to another.

No one said much about why at home. At school
Jenny told me in her Frankenstein whisper,
'He was gay!'

Gay? And wasn't it a good loving thing
to be gay? As good as to be kind
I thought then,

and said as much to my silent mother
as she wrung out a cloth until her knuckles shone,
white bone under raw thin skin.

The Musician's Widow

Plants she loved, all growing things.
Soil was nourishment richer than food for her,
richer than sex as she grew older. But she
hated death, hated his unjust death in particular.
The music of him tunnelled through her mind to pursue her.

For a time she remembered to return to him.
She left the upholstery of the new home, the ferns,
the fuchsias, the piano-furnished living room,
and followed her spade into the cold warmth
his absence had hollowed out for her in earth.

Desire for him she burned with his body, though.
Detritus of nostalgia was a waste of good.
And she had to come back as woman to the world,
as a green branched seedling of her purpose, need—
the life behind her gaping like a seed.

The Road In

For Edward Lowbury

Driving white streets
through the black city,
a city built white over black
where the people are lonely,
where night is blonde as ignorance
under the arc lamps, where
the way will not easily be found
in the star-bare darkness, where
tall racks of candles
gutter into glass without heat, where
everyman, climbing his footsteps
to eyes on the fourteenth storey,
lies in the desert of the moon
to wake or to sleep; no wonder

we are lost and quarrelling
in this future of streets
that will not believe in the door
we are looking for, or
in porchlight and leaf-pattern
thrown out alive on the lawn, or
in lilacs poured over a fence, or
in one man, backed
by his books and a piano, smiling
at the city around him, while
upstairs someone can be heard
playing Bach on the virginals,
and someone else pretty is laughing,
tuning a cello.

A Dream of Stones

For Norman Nicholson

I dreamed a summer's labour,
loss or discovery,
had brought me, on the sand,
to a nest of stones.
What shall I do with these stones
that shine too weakly to be gems,
that might be seeds?

Stones are to build with,
but here there has to be
sand to the sea, bare land and sea.

Why, since these stones
are moist and heavy with sacrifice,
should they not be planted?
There are no trees here.
Maybe there are trees
coiled in the wilderness
of the stone seeds.

I am pocking the soil with my heel.
Here, here, here, here.
Into each footprint, a glimmering pearl.

They will not be counted,
these seeds, these stones, these
possible offerings from impossible language.
They resist being tears.
I tell them to you now as if they were things
too alive to be left unburied
under common years.

Making Poetry

'You have to inhabit poetry
if you want to make it.'

And what's 'to inhabit'?

To be in the habit of, to wear
words, sitting in the plainest light,
in the silk of morning, in the shoe of night;
a feeling, bare and frondish in surprising air;
familiar . . . rare.

And what's 'to make'?

To be and to become words' passing
weather; to serve a girl on terrible
terms, embark on voyages over voices,
evade the ego-hill, the misery-well,
the siren hiss of *publish, success, publish,
success, success, success.*

And why inhabit, make, inherit poetry?

Oh, it's the shared comedy of the worst
blessed; the sound leading the hand;
a wordlife running from mind to mind
through the washed rooms of the simple senses;
one of those haunted, undefendable, unpoetic
crosses we have to find.

In the Tunnel of Summers

Moving from day into day
I don't know how,
eating these plums now
this morning for breakfast
tasting of childhood's
mouth-pucker tartness,
watching the broad light
seed in the fences,
honey of barley,
gold ocean, grasses,
as the tunnel of summers,
of nothing but summers,
opens again
in my travelling senses.

I am eight and eighteen and eighty
all the Augusts of my day.

Why should I be, I be
more than another?
brown foot in sandal,
burnt palm on flaked clay,
flesh under waterfall
baubled in strong spray,
blood on the stubble
of fly-sweet hay—
why not my mother's, my
grandmother's ankle
hurting as harvest hurts
thistle and animal?
a needle of burning,
why this way or that way?

They are already building the long straw cemetery
where my granddaughter's daughter has been born
    and buried.

Poem for Harry Fainlight

(d. 1982)

Tree—a silence
voiced by wind.

Wind—breath
with a tree's body.

Axe the bole.
Plane the boards.

Here is Art,
the polished instrument,
casket and corpse.

Dum Vixi Tacui
Mortua Dulce Cano*

The harp's motto
will do for the harpist's apology.

But your poems, Harry,
those Welsh oaks
stunted by the wind's scream?

They were always
transforming
your wrong life
into their live
silence . . .

* While I lived I was silent.
  In death I sweetly sing.

Red Rock Fault

For Frances Horovitz (d. 1983)

This is the South-West wind
the North-East breathes and knows,
that lifts linoleum under kitchen doors,
that bends thorned trees one way on the moors,
that hooks back little white knots of the Irthing
in shaggy impermanent weirs
by the empty farm at the river's turning
where spiders make nets for the silted windows
and machinery rusts in the byres.
Fran, has it been two years?

I see you again in your boy's coat
on that sudden and slithery hill of stones
where we ducked from the wind one afternoon
when slant light cut and shone
through glass-white arcs of October grass.
It was just by the Red Rock Fault
where limestone meets sandstone, lass.
You carried your love of that rushy place
in the candle of your living face
to set in the dark of your poems.

And now we have only the poems.
And the snow-light, water-light winter still
will come to that ridge of Roman stones
by Spadeadam, Birdoswald, high Whin Sill
where so few trees lose uncountable leaves
to this wind—one breath from uncountable lives.
Shrill clouds of gathering jackdaws, starlings
storm an enormous sky.
That hugh split ash by the ruined steading—
Cocidius, life-keeper, live eye.

Willow Song

I went down to the railway
But the railway wasn't there.
A long scar lay across the waste
Bound up with vetch and maidenhair
And birdsfoot trefoils everywhere.
But the clover and the sweet hay,
The cranesbill and the yarrow
Were as nothing to the rose bay
    the rose bay, the rose bay,
As nothing to the rose bay willow.

I went down to the river
But the river wasn't there.
A hill of slag lay in its course
With pennycress and cocklebur
And thistles bristling with fur.
But ragweed, dock and bitter may
And hawkbit in the hollow
Were as nothing to the rose bay,
    the rose bay, the rose bay,
As nothing to the rose bay willow.

I went down to find my love,
My sweet love wasn't there.
A shadow stole into her place
And spoiled the loosestrife of her hair
And counselled me to pick despair.
Old elder and young honesty
Turned ashen, but their sorrow
Was as nothing to the rose bay
    the rose bay, the rose bay,
As nothing to the rose bay willow.

O I remember summer
When the hemlock was in leaf.
The sudden poppies by the path
Were little pools of crimson grief,
Sick henbane cowered like a thief.
But self-heal sprang up in her way,
And mignonette's light yellow,
To flourish with the rose bay,
    the rose bay, the rose bay,
To flourish with the rose bay willow.

Its flames took all the wasteland
And all the river's silt,
But as my dear grew thin and grey
They turned as white as salt or milk.
Great purples withered out of guilt,
And bright weeds blew away
In cloudy wreathes of summer snow,
And the first one was the rose bay,
    the rose bay, the rose bay,
The first one was the rose bay willow.

Dreaming of the Dead

For Anne Pennington (d. 1981)

I believe, but what is belief?

I receive the forbidden dead.
They appear in the mirrors of asleep
To accuse or be comforted.

All the selves of myself they keep,
From a bodiless time arrive,
Retaining in face and shape

Shifting lineaments of alive.
So whatever it is you are,
Dear Anne, bent smilingly grave

Over wine glasses filled by your fire,
Is the whole of your life you gave
To our fictions of what you were.

Not a shadow of you can save
These logs that crackle with light,
Or this smoky image I have—

Your face at the foot of a flight
Of wrought-iron circular stairs.
I am climbing alone in the night

Among stabbing, unmerciful flares.
Oh, I am what I see and know,
But no other solid thing's there

Except for the terrible glow
Of your face and its quiet belief,
Light wood ash falling like snow

On my weaker grief.

BLACK GRATE POEMS

For Paul Winstanley and the former
inhabitants of his house and our house

November

All saints and all souls,
martyrdom of the good days.
Daylight is smoke out of the dark's bonfire.
An old sun huddles in unclean caves.
But here, anyway, is this step,
now another step.
In imaginary fields, a tractor
sputters with purposes.
As black coal in our black grate
ignites in uncertain tongues,
birch-blaze thins over clinker
where the coke works were.

Household Gods

The room is silent except for the two hearth spirits.

The fire speaks out of the grate like a kindly tongue.
The man speaks out of the square screen like a god.

The fire burns slowly, holding itself back from burning.
The man speaks quickly, hurtling himself into particles.

Hold up your hands to the fire, and they, too, are fires.
Hold your hands up to the screen, and feel the premises of
    illusion.

Wherever you move, the fire pulls you close like a magnet.
Wherever you look, the screen intercepts your escape.

When the fire is worshipped, the resident cats will pray with
    you.
When the screen presides, it lashes the dog with its scream.

The fire has nothing to tell you; it waits for your thoughts.
The screen has to tell you everything except what you are.

In heaven, they will give you to fire to be consumed into
    freedom.
In hell, they will play you over and over on the tape of your
    dead life.

In hell, nothing you have done will be not watched.

Demolition

They have blown up the old brick bridge
connecting the coal works with the coke works.
Useful and unimposing,
it was ever a chapel of small waters,
a graceful arch toothworked with
yellow bricks notched into red bricks,
reflecting there sudden bright winks
from the Browney—an oval asymmetrical image
which must have delighted, as fisher-children,
these shiftless but solid grey men
who follow so closely the toil of its demolition.

The digger's head drops and grates and swings up,
yellow fangs slavering rubble and purple brickdust;
but the watchers wear the same grave, equivocal expression.
They might be grieving
(their fathers built it, or their fathers' fathers)
or they might be meaning
Boys won't be going to the mine no more.
Best do away with what's not needed.
That's Jock Munsey's lad in the cab there, surely.
Good job it's at home, not away on the telly.

February: Track to Lanchester

The Black Burn rattles into the brown Browney,
but the dominant colour of the season is silver,

bright tin of bare beechwood in light
diffused white out of one shallow cloud,

pewter at the heavier snow-black horizon where
intense hills—studded here, there, with live pylons—

ignite nearer railings, wires, roofs, windy larches,

between which, faint-breathing, much warmer mud
persuades rosy rags out of vertical birches.

Confusion of voices. Dispersal of finches.

The track, insubstantial as stone over mercury,
does really hold us, means to support us.

As it does those on motorbikes behind and before us
whose brown sputter throws up black grit in their faces.

Spring Song
of the Poet-Housewife

The sun is warm,
and the house in the sun
is filthy . . .

grime like a permanent fog
    on the soot-framed windowpanes,
dust, imprinted with cat's feet,
    on the lid of the hi-fi,
dishes on the dresser
    in a deepening plush of disuse,
books on the blackened shelves,
    bearing in the cusps of their pages
    a stripe of mourning . . .

The sun is warm,
the dust motes and dust mice
are dancing,

the ivies are pushing green tongues
    from their charcoal tentacles,
the fire is reduced to a
    smoky lamp in a cave.
Soon it will be spring, sweet spring,
    and I will take pleasure in spending
many hours and days out of doors,
    away from the chores and bores
    of these filthy things.

Signs

A green cross on the trumpet lip of the snowdrop.
Shard of a red pot scuffed by a startled jay.
Smoke like visible wind where T.V. antennae
Root in suburban chimneys and shining day.

A child again, my beginning again waking up
After fifty years of change through the same circle.
By light unnerved in this eggshell air of April,
Sharp, clean, chill to the astonished nostril.

Gales

Weather buffets our houses in armour all night.
There is no sleep that is not a war with sound.
Morning is safer. A good view of the battleground.
And no roof that is not between light and light.

A Prayer to Live with Real People

Let me not live, ever, without fat people,
the marshmallow flesh set thick on the muscular bone,
the silk white perms of sweet sixteen-stone ladies,
luscious as pom-poms or full blown perfumed magnolias,
with breasts like cottage loaves dropped into lace-knit sweaters,
all cream-bun arms and bottoms in sticky leathers.
O Russian dolls, o range of hills
rosy behind the glo-green park of the pool table,
thorns are not neater or sharper than your delicate shoes.

Let me not live, ever, without pub people,
the tattooed forearm steering the cue like a pencil,
the twelve-pint belly who adds up the scores in his head,
the wiry owner of whippets, the keeper of ferrets,
thin wives who suffer, who are silent, who talk with their eyes,
the girl who's discovered that sex is for she who tries.
O zebra blouse, o vampish back
blown like a lily from the swaying stalk of your skirt,
roses are not more ruthless than your silver-pink lipstick.

Let me live always and forever among neighbours like these
who order their year by the dates of the leek competitions,
who care sacrificially for Jack Russell terriers and pigeons,
who read very carefully captions in *The Advertiser* and *Echo*
which record their successes and successes of teams they
    support,
whose daughters grow up and marry friends' boys from Crook.
O wedding gifts, o porcelain flowers
twined on their vases under the lacelip curtains,
save me from Habitat and snobbery and too damn much
    literary ambition!

Pit April

Coal fires
that have not gone out all winter
still stoke brown air
with soot and smoky clothes.

On sagging staves of wires
between low poles
quavers and crotchets are jackdaws,
huge crows.

Wasteground resounds with caws
and two-stroke cackles; here's
haloed and scaly coltsfoot,
pussy willow almost
hedgehog willow, north April

all out in pit-holes
frothy with yellows—
aconite, primrose, gorse,
celandine? . . . cellophane.
Daffodils in allotments
and lacy windows.

The always adaptable dandelions'
brassy screws
batten down tips with topsoils'
richer intentions.

And now a summer opens its healed views
of particularly green loyal
undulating furrows.

Forgotten of the Foot

Equisetum, horsetail, railway weed
Laid down in the unconscious of the hills;
three hundred million years' dream still buried

In this hairsoft surviving growth that kills
Everything in the glorious garden except itself,
That thrives on starvation, and distills

Black diamonds, the carboniferous shelf,
That was life before the animals,
Before trilobites and shellfish,

A stratum of compressed time that tells
Truth without language, is the bodystore
Of fire, light, a night without intervals,

That becomes people's living only when strange air
Fills out the folded lungs, the inert corpuscles . . .
Into the mute dark, loud life once more.

> *Proggie mat, proggie mat,*
> *Who will mend my proggie mat?*
> *Lay it down and squash it flat*
> *And find your knickers after that.*

★

So the hills must be pillaged and cored.
Such evidence as they hide must be hacked out
Urgent as money, the buried black seams uncovered.

Rows of stunted houses under the smoke,
Sootblack houses pressed back hard against pit,
By fog, by smoke, by the cobra hood of smouldering coke

Swayed from the nest of ovens huddled opposite.
Families, seven or ten to a household,
Growing up, breathing it, becoming it.

On winter mornings, grey capped men in the cold,
Clatter of boots on tarmac, sharp and empty,
First shift out and thick frost simple as gold

On the sulphurous roofs, on the stilted gantry
Crossing to engine house and winding gear.
Helmet, pick, lamp, tin bottle of tea.

> *Night by day and*
> *Night by night.*
> *Bituminous night*
> *Is the miner's light.*

★

A Nan or Nora servant to each black grate.
Washing on Monday, water kept warm in its well.
Iron and clean on Tuesday, roll out and bake

Each Wednesday, with that cokeish hot bread-smell
No child who grew up here forgets.
Thursdays, the Union and the Methodist circle.

Fishday on Friday (fryday), a queue of kids,
Thin, squabbling by the chippy. The men out for pool
After pay day. Later, swearings and broken heads.

Wheels within wheels, an England of working Ezekiels,
Black dangerous water boiling in pits, for cooling,
Forges roaring and reddening, the black irons
    glowing like jewels.

No more, no more. They've swept up the workings
As if they were never meant to be part of memory.
A made way of being. A working place. Living a living.

> *You get the fish,*
> *I get the bone,*
> *You get the apricot*
> *I get the stone.*

★

Prim Esh looks down on the red-tiled brick town's soul
Streaming from its roofs in the smoke of a lost century—
A veil of breath in which to survive the cold.

When the mine's shut down, habits prolong the story,
Habits and voices, till grandmothers and old ways pass,
And the terraces fold back into themselves, so black, ugly

And unloved that all but the saved (success
Has spared them, and the angel of death-by-money) move
   away.
The town is inhabited by an alien, washed up innocence.

Children and animals and people too poor to stay
Anywhere else stray, dazed, into this slum of Eden.
The church is without glass saints or statuary.

The memorial is a pick, a hammer, a shovel, given
By the men of Harvey Seam and Victoria Seam. May
Their good bones wake in the living seams of heaven.

> *He breaketh open a shaft away from where men sojourn.*
> *They are forgotten of the foot that passeth by.*

RHYMINGS

Claude Glass

For Robert and Pamela Woof

Eyes are too close to Nature to be nice,
So Claude's disciples thought of a device
Through which they could evade the messy world
By catching it in image as it curled
Within a glass held up before its face
To give God's barbarous hills and rivers grace.

His name became an impulse to impart
To Nature all *les belles finesses* of Art;
Taught British tourists for a century
To turn their backs on what they went to see.
Meanwhile the men of coal and iron and steel
Took out of Nature what they knew was real.

Now from the Tyne's black stacks and blacker steam
We drive out to the Lakes and their museum
To seek what never was but always risked
The truth to be the fair, the picturesque—
A landscape-cognac that the connoisseur
Will recognize—expensive, sweetened, sure.

Re-reading Jane

The memorial to Jane Austen in Winchester Cathedral reads:

In memory of JANE AUSTEN, youngest daughter of the late Revd George Austen, formerly rector of Steventon in this county. She departed this life on the 18th July, 1817, after a long illness supported with the patience and the hopes of a Christian. The benevolence of her heart, the sweetness of her temper, and the extraordinary endowments of her mind obtained the regard of all who knew her and the warmest love of her intimate connections. Their grief is in proportion to their affection. They know their loss to be irreparable, but in their deepest affliction they are consoled by a firm though humble hope that her charity, devotion, faith and purity have rendered her soul acceptable in the sight of her REDEEMER.

To women in contemporary voice and dislocation
she is closely invisible, almost an annoyance.
Why do we turn to her sampler squares for solace?
Nothing she saw was free of snobbery or class.
Yet the needlework of those needle eyes . . .
We are pricked to tears by the justice of her violence:
Emma on Box Hill, rude to poor Miss Bates,
by Mr. Knightley's *were she your equal in situation—*
but consider how far this is from being the case
shamed into compassion, and in shame, a grace.

Or wicked Wickham and selfish pretty Willoughby,
their vice, pure avarice which, displacing love,
defiled the honour marriages should be made of.
She punished them with very silly wives.
Novels of manners? Hymeneal theology!
Six little circles of hell, with attendant humours.
For what do we live but to make sport for our neighbours
And laugh at them in our turn? The philosophy
paused at the door of Mr. Bennet's century;
The Garden of Eden's still there in the grounds of Pemberley.

The amazing epitaph's "benevolence of heart"
precedes "the extraordinary endowments of her mind"
and would have pleased her, who was not unkind.
Dear votary of order, sense, clear art
and irresistible fun, please pitch our lives
outside self-pity we have wrapped them in,
and show us how absurd we'd look to you.
You knew the mischief poetry could do.
Yet when Anne Elliot spoke of *its misfortune*
to be seldom safely enjoyed by those who
enjoyed it completely, she spoke for you.

Epitaph for a Good Mouser

Take, Lord, this soul of furred unblemished worth,
The sum of all I loved and caught on earth.
Quick was my holy purpose and my cause.
I die into the mercy of thy claws.

Divorcing

After Gertrude Stein

I am I because my little dog knows me.
We are we because our little dog knows us.

I am I, but my little dog knows you.
You are you, but your little dog knows me.

I am I. You are you.
Poor little dog. Poor little dog.

On Watching a Cold Woman
Wade into a Cold Sea

The way that wintry woman
walked into the sea
was as if, in adultery,
she strode to her leman.

Something in the way she
shrugged off her daughters'
moping by the sea's hem
as if they were human

but she of the pedigree
and breed of Poseidon,
slicing through the breakers
with her gold plated knees,
twisting up her hair
with a Medusan gesture,

something in the augury
she shook from her nature
made women look at women
over stiff cups of tea,
and husbands in their season
sigh suburbanly to see her.

O go dally with your children
or your dogs, naked sirs;
the venom of the ocean
is as kindness to hers.

The Parson and the Romany

A Black Mountain Ballad to a Green Mountain tune

The parson went out one stormy day
To visit the sick in his valley grey.
A Romany girl he met on his way
With eyes like the radiant dawn of day,
　　And she lived in the weather all around, all around,
　　And she lived in the weather all around.

'O tell me, parson my love,' she said,
'Where are you going with your sickle head,
And your long black cloak from your shoulder spread,
And your stoop like a monument to the dead,
　　When you live with the weather all around, all around,
　　When you live with the weather all around?

'When you are as old as I, my lass,
You'll learn how the hard years press and pass,
For grief is the text and pain is the task,
And everyone belongs to the weeping class
　　While we live in the weather all around, all around,
　　While we live in the weather all around.'

'And what will you do when your time is done,
And you count up your sufferings one by one,
And put them in a sack with the string undone
For inspection by the Lord and his Ghost and Son
　　As we pray in the weather all around, all around,
　　As you pray in the weather all around?'

'Are you a demon or a sprite,' he cried,
'Do you speak out of ignorance or of pride?
Look up where the clouds are gaping wide
To show you the pillars of Hell inside,
　　While you laugh in the weather all around, all around,
　　While you laugh in the weather all around.'

'That's odd,' said the girl, 'for I've just come
From an angel who was sitting in a pillar of the sun.
He blessed me and called me his chosen one
And soon I'll be having a pretty little son
   Who will live in the weather all around, all around,
   Who will live in the weather all around.'

The parson, he shook off his cloak and hood,
He threw back his head and he laughed where he stood.
'Now tell me your name, O my wicked and good.'
'They call me the Lady of Kilpeck wood.'
   So they danced in the weather of the sun, of the sun,
   So they danced in the weather of the sun.
   They danced in the weather of the sun all around,
   And they danced in the weather of the sun.

A LEGACY
On my Fiftieth Birthday

After François Villon

In nineteen hundred, eighty-three,
At fifty, young in wisdom though
Too old to quarrel with poverty,
Unwilling to (unscripted) go
Into the wilderness I know
Will give me grit to write as if
Love were the object, not the show
Of reputation in this life.

I, in my fathers' name, profess
Belief in music, cadence, rhymes,
And to the present mob confess
Indebtedness to simpler times—
To Chaucer's tales and Villon's crimes,
To sacred fire that Herbert wrought
Into his sacrificial lines,
To Yeats's Irish argonaut.

And in my mothers' memory
I praise that panoply of wit
Which turned, in style, on tyranny
(Like Beatrice on Benedick)
The wily cheek that conquered it.
Two Emilies I praise, George, Jane,
And from the Sixties' squalid pit,
Elizabeth who kept me sane.

Now in this season of the goat
Who stamps my passage with his sign,
When the wind howls from its wolf's throat,
And windows weep with ice or rain,
Let me set down in words as plain
As Villon's stanza will allow
The legacy that in my name
I'd like to leave the world I know.

Firstly, my heart—which signifies
All that I've loved and, loving, wish
To salvage from this hive of lies
I've lived in since ambition kissed
Its sting into my quietness—
To Michael Farley, poet-fool,
Who represents the best, undressed
Conditions of our compound soul.

To Andrew Motion, any fame
That to my thin books may adhere
Like sheep's wool to a barbed-wire name.
Further, my reputation for
High-flown frigidity to Fleur
Who in the Oxford lists has been
The lady aptest to prefer
The acid to the saccharine.

To Raine and Reid (the Martian lot)
My recipe for onion soup,
That fresh peas may be spared the pot
And treated as befits the group
Pisum sativum, well brought up.
With simple, short, three-minute heating,
Served *au beurre*, but *sans trop* trope,
A pea can be delicious eating.

Beneath its Gutenbergs there snore
Eurekas every onion stows
In layers and layers of metaphor;
You boil them out like stains from clothes
Before you add the salt and Beauj-
Olais. But why should I repeat
What every kitchen poet knows?
Spice well, and throw away the meat.

And, since the boys have set up court
So quaintly in the scullery,
We girls, why not? may now depart
To sit in the academy.
Here is the grill, the sink, the tea,
The colander to catch the dregs.
And here's (for an emergency)
That thing we use to time the eggs.

But wait. I'll turn and name the rich
Accumulations of my plate.
From Donald Hall (Ann Arbor, Mich.)
I learned my trade. To him, too late,
I leave my green, inebriate
Midwestern anxiousness to please,
My appetite for appetite,
My love affair with Lymeswold cheese.

To Glasgow, now I've left the place,
I leave a midden of regrets,
A name I should have worn with grace
Stubbed out in bitter cigarettes.
No poem can rectify the debts
My madness, mixed with alcohol,
Incurred in sundry tenements
Anent the academic brawl.

And yet, if I can make my peace
With Philip Hobsbaum, in whose book
No Structuralism found increase
Nor foolish Hermeneutic took
The liberty of playing hook
To literary accolades—
Well, then I will. He never took
His flag down from the barricades.

Yet, if I had to say which were
The saddest of my salad years,
The time I'd give back, with a purr,
To some great *Lanark* in the spheres—
I think I'd weep Glaswegian tears
For Tom and Liz and Angus, whose
Subversive talents warmed their beers
And talked the curse off college booze.

By Tentsmuir's Tayport, where the Tay
Spills out in salty, spatulate
Redundancies of tidal clay,
I buried all that out-of-date
Hysteria of want and hate.
In Fife I count among my friends
The spumey bay, the slanted light—
Ablutions for us puritans.

With Geoffrey Dutton, Alan Wall,
Anne, Ellie, Monty, mild Bill Tait
Who outdrank and out-owled us all,
Those *Seagate* days I celebrate
From David's Fort to Nethergate.
Down here I can't use all my vice;
I'll leave you half, in case some fate
Decides Dundee deserves me twice.

To Douglas Dunn, who was a Scot
Before he was a poet—eyes
To see that what the North East's got
(Without a knack for compromise)
Is pride, deep humour—and those sties
Its desperate young consider right
To fight inside, or patronize.
Notice Orion, though, at night,

Or watch the shelduck, dunlin, terns
Perform their ritual antic dance
Where water meets the sand and churns
On every beachy prominence
Topographies of innocence.
You call me a Romantic? I'm
Too old to frown or take offence.
We live in a diminished time.

The minor English sip their verse
From sherry glasses as they talk.
Fearful of spirit, they endorse
The safe and unpretentious bloke
Who slips a knowing little joke
Betwen big gulps of mum and dad
And nips of sweetened back-yard folk.
I wish the outlook weren't so sad.

Something in poetry goes wrong
When poets tacitly agree
There's nothing more to say in song.
Our journalists' and linguists' plea
For unrestricted novelty
Shores up the bits, but shreds the whole.
O where is that great-rooted tree
Yeats made a symbol of the soul?

To Peter and Penelope
Whose wise, outlandish confidence
In holiness and witchery
Upbraids the Positivist sense
That's peddled in the *TLS*,
I leave one caution: Satan makes
A plausive honey in his nest;
Don't treat him like a friend of Blake's.

Near Hay I think Traherne and Vaughan
Are angels in that border air.
My jealous, butting Oxford tongue
Dried up while I was living there.
My *pauvre âme* began to tear
The bars down from inside its cage,
And black theatrical despair
Rose like a curtain from a stage.

I don't mean evil isn't real.
Dear God, things hardly could be worse.
The tragedy is that we feel
Important when we preach in verse
Or march to mitigate the curse
Of mass greed, hatred and the bomb.
We fear a vacant universe,
Yet Yeats's Chinamen were calm.

To Roger Garfitt, all my strolls
Along the Wye—to write his book.
(To Flash and Spark, the rabbit holes
Poor Guinness wistfully forsook.)
Glenn Storhaug, once you undertook
To speak in printing, you became
So indisputable you put
Commercial publishing to shame.

From every proof you pull, I learn.
Dear friend, I leave you and your press
This deep-sworn promise to return.
To Alan Halsey, too, success
For poetry. And happiness
In Broad Street's number twenty-two.
(I hope we never have to mess
With real estate again, don't you?)

To John, my son, who at sixteen
Bids fair to beat a meaner drum
Than any public star I've seen,
I leave the music I've become
Too deaf to hear or profit from.
I'd give him every hour I've known
Of Mozart, Schubert, Beethoven,
If he could give me back one tune.

You, Charles, must take what voice I can
Dredge up from years of broken rules.
Licence the beat, or let it scan,
But shun the literary fools
Whose verses reek of cliques and schools;
Nor let the poet fight the heart.
The only clean and honest tools
Are truth, good whisky, and good art . . .

And love, of course, which at the start
I meant to make the heroine
Of this homage *The Devil's Fart*
Set off in France. Ah, *cher Villon*,
They tell me that my woman's tongue
Must dredge my womb to find its root,
That verses in the masculine
Subvert the Female Absolute.

I'd rather be a pagan sucked
At some outlandish creed, absurd,
Than be indubitably fucked
And have to find another word.
I don't like 'poet's moll' or 'bird',
But 'chauvinist' and 'sexist war'—
Dragged in to keep the anger stirred—
Are just as twisted at the core.

This haunts me—this profound belief
That what's between us and you males
In this Anatomy of Grief
Is *Selfishness* in all its veils.
I think of Gillian in Wales,
Of Jeremy, of Robert Wells,
Of Frances Horovitz's trials;
God spare them equally my hells.

To Caroline, my daughter, who's
A quarter century to my half,
I leave my hard-won stoic views,
My silliness, to make her laugh;
Also this three-line epitaph:
Here lies a mother who, in flame
Of life, lost all its grain. O chaff
Be charitable to her name.

To my good and loyal Guinness, this:
A bouquet of assorted sticks
That connoisseur of canine piss
Can take on wet, olfactory walks.
To both my cats, a furry box
Of heart-and-kidney flavoured spice.
Also my conscience, so those crooks
Won't catch my friends, the birds and mice.

To all my students in this age
Of terror masked as arrogance
When self-regarding verbiage
Is mostly personal defence,
I leave my tender deference
To poets older than the bomb
Who temper grief with assonance
And wise, if dearly paid for, calm.

Still thriving in the English rain
Are Annes the Oxford angels keep.
Both Peter Levi and John Wain
Can tell a poet from a sheep.
If Yeats and Dylan from their sleep
Could rise, with Auden, to renew
The talk they wanted to repeat,
John Heath-Stubbs, they would come to you.

To Geoffrey Hill, awe and applause
For your great homage to Péguy.
You wring from tight prosodic laws
Such music, such profundity
The moving words forget to be
Pieces of language and become
Sacred as tools. So charity
Imbues with grace your finest poem.

To Tom, a river full of praise
That he may fish and find his trout
Symbolic, simple and ablaze
With cleansing blood to write about.
My Oxford, though, is not without
Its unwashed corners, black with much
Suspicion, prejudice and doubt . . .
Wounds still too resident to touch.

Cambridge that gave me birth and name,
Magnet, unmerciful and strong,
In which I found girl-love, grown-shame,
In which I'll die, unless I'm wrong
About where human souls belong.
Twice you have been my home, but three
Is the number drawn and drawn
In my self-casting constantly.

And so, as Lady Memory
Undoes the clasps of her *armoire*,
Unfolds my soiled identity
And, piece by piece, my repertoire
Of gross mistakes—by which we are
Defined and moulded by the Muse—
I thank her that I've come this far
And have so little left to lose.

If I were Berryman I'd swear
That Villon visited my sleep.
But I'm no Bellë Heaulmière.
I don't think I'd be Villon's *type*.
Nevertheless, I think some deep
Affinity of drink or rhyme
(Plus how we rarely earn our keep)
Links poets in a ring of time.

The best of everything the world
Affords, and therefore coins as grace—
Success etc.—is curtailed
By wild defects of sex and race.
The devil's fart is from his face.
I praise some *patria mea*—odd,
For it's a state but not a place—
I call it 'Listening for God.'

As for my eyes, my ears, my teeth,
The little lusts that live therein,
They can dissolve like salt beneath
The ink and paper of my skin.
For now and since the glacier's been
The boulder clay brings down the stones.
The tide purls out, the tide purls in.
So may a white sea wash my bones.

Notes

St. 3 Elizabeth = Elizabeth Bishop
     5 poet-fool = for the sense of 'fool' see Charles Williams's novel
        The Greater Trumps
     6 Fleur = Fleur Adcock
    13 *Lanark* = the novel of that title by Alisdair Gray
    13 Tom = Tom Leonard
    13 Liz = Liz Lochead
    13 Angus = Angus Nicholson
    20 Peter and Penelope = Peter Redgrove and Penelope Shuttle
    23 Flash and Spark = Roger Garfitt's lurcher dogs
    23 Guinness = my labrador x collie dog (also st. 31)
    26 Charles = my younger son
    29 Gillian = Gillian Clarke
    29 Jeremy = Jeremy Hooker
    33 Annes = Anne Ridler, Anne Born and Anne Pennington
    35 Tom = Tom Rawling